Firewalla: After Setup

An Operator's Guide to Running a Secure Home Network

L. Phillips

PRAIRIE LIGHT PRESS
ESTD. 2025

Helping readers move from the unknown to the known

Published by Prairie Light Press through IngramSpark

Publication Identifiers:

ISBN (Hardback Case Laminate) 9798996980246
ISBN (Paperback) 979-8278933922
ISBN (Kindle) B0G765KDBT

Fifth Edition

Disclaimer

The information in this book is provided for educational and informational purposes only. While every effort has been made to ensure accuracy, technology and security practices evolve over time.

The author makes no guarantees regarding the effectiveness of any tools, techniques, or strategies discussed and assumes no responsibility for errors, omissions, or outcomes resulting from the use of this information.

Readers are encouraged to evaluate their own circumstances and, where appropriate, consult qualified professionals before implementing security, financial, or technical decisions.

Non-Affiliation Notice

This book is an independent publication and is not affiliated with, endorsed by, or sponsored by any company or organization mentioned herein, including but not limited to Firewalla.

Get Started Right Away with our Quick-Start Guide on page 1

Table of Contents

QUICK-START FOR OPERATORS (90 MINUTES)

This book is designed to be read end-to-end, but you don't need to do everything at once.

If you have about **90 minutes**, this section gives you the highest-security return for your time. You can come back later for depth.

0–10 MINUTES: CLARITY

- Read **Introduction** to adopt the operator mindset
- Skim **Chapter 1** to understand device discovery
- Ensure most devices are identified and named (perfection not required)

Goal: You know what's on your network and why.

10–30 MINUTES: STRUCTURE

- Read **Chapter 2** (Naming Conventions)
- Read **Chapter 3** (Grouping Strategy)

Apply immediately:

- Use a consistent naming convention
- Create at least **Trusted**, **IoT**, and **Guest** groups
- Assign every device to a group

Goal: You can reason about your network at a system level.

30–45 MINUTES: BASELINE PROTECTION

- Read **Chapter 4** (Built-In Protections)

Enable:

- Active Protect (Default mode)
- Ad Block (network-wide or IoT-only)

Goal: To remove ambient risk without ongoing effort.

45–70 MINUTES: CONTAINMENT

- Read **Chapter 5** (One-Way Communication)
- Read **Chapter 6** (Blocking Lateral Movement)

Apply:

- Block IoT devices from initiating connections to local networks
- Block IoT-to-IoT communication

Goal: Prevent small compromises from spreading.

70–90 MINUTES: EXPOSURE REDUCTION & HABITS

- Skim **Chapter 7** (Region & Country Blocking)
- Read **Chapter 8** (Operational Awareness)

Apply:

- Add conservative geo-blocking for IoT or Guest groups

- Establish a simple weekly review habit

Goal: Reduce noise and make problems obvious.

WHAT TO SKIP FOR NOW

- Fine-grained exceptions
- Advanced tuning
- Edge-case optimizations

Those come later, once the system is stable.

QUICK-START TAKEAWAY

If you complete this path, your network will be:

- Structured
- Contained
- Quiet
- Sustainable

You can now read the rest of the book for depth instead of urgency.

NOTE TO READER

This edition has been refined for clarity, pacing, and long-term usability.

No concepts were simplified. No security posture was weakened.

You'll notice:

- Cleaner sectioning for scan-reading
- Consistent terminology and tone
- Priority callouts where decision overload is common
- Operator Takeaways that reinforce judgment, not checklists

This is an operational manual. Read actively.

Disclaimer

This book provides general guidance based on common home-network environments and typical Firewalla configurations. Every network is different.

You are responsible for understanding your own environment, testing changes incrementally, and adjusting rules to fit your specific needs.

No security configuration is risk-free, and no guidance replaces informed judgment.

INTRODUCTION — BEYOND BASIC SETUP

So you've got Firewalla installed. It's running. The lights are blinking. Devices are connecting. Internet works.

Congratulations—you've completed the easy part.

Now comes the real work: actually using Firewalla effectively. This is where most people struggle. Not because Firewalla is difficult, but because network management requires ongoing attention, decision-making, and refinement. It's not a set-it-and-forget-it appliance.

This book is about that operational phase—the weeks, months, and years after setup. It's about the daily decisions, the maintenance routines, and the judgment calls that determine whether Firewalla becomes an essential security tool or just another box gathering dust.

WHAT THIS BOOK IS (AND ISN'T)

This is not a setup guide. We assume Firewalla is already running in Router Mode with devices connected. Installation, onboarding, and feature tours are deliberately excluded.

This is an operational manual.

Specifically, this book focuses on:

- Making naming decisions that scale
- Organizing devices for long-term management
- Creating rules that survive real-world use
- Recognizing what is normal vs what deserves attention

- Building habits that keep your network secure without constant babysitting

Think of it this way: the setup guide gave you a working car. This book teaches you how to drive it well, maintain it properly, and avoid common accidents.

WHO THIS BOOK IS FOR

You're the intended reader if:

Firewalla is installed and working

- Your device list is cluttered or confusing
- You ignore most alarms because there are too many
- You've created rules that don't seem to help
- You want better security but aren't sure what to *do*

You'll get the most value if you're willing to invest a few focused hours up front. Organization and clarity reduce ongoing effort dramatically.

THE OPERATIONAL MINDSET

Effective network management requires a mindset shift.

Think in systems, not one-off fixes. When something goes wrong, ask whether it represents a pattern that deserves a structural response.

Embrace iteration. Your configuration will evolve. That's normal.

Balance security and usability. Over-restriction leads to workarounds that undermine security entirely.

Invest time early to save time later. Naming and grouping may feel tedious, but they pay dividends forever.

Document decisions. Future-you will not remember why a rule exists.

Identification works best as a loop. Observe first, name devices once you're confident, and revisit periodically as behavior and inventory change.

OPERATOR TAKEAWAYS

- Firewalla effectiveness depends on *how it's operated*, not how it's installed
- This book is about judgment, habits, and systems—not features
- You're adopting an operator mindset, not a consumer mindset

NEXT: We stop guessing and start identifying what's actually on your network.

CHAPTER 1 DEVICE DISCOVERY & IDENTIFICATION

Before you can secure your network, you need to know what's on it.

Most networks begin in partial ignorance: dozens of devices, many unnamed, some unknown. This chapter turns that uncertainty into inventory.

IDENTIFICATION IS A PROCESS, NOT A TASK

Device discovery is not something you finish once. New devices appear. Old ones disappear. Patterns emerge over time.

Treat identification as an investigative process rather than a cleanup chore. This mindset prevents frustration and encourages iteration.

Enable Active Protect first, then turn on Ad Block. Family Protect can be added later once structure and visibility are stable.

IF YOU ONLY DO ONE THING

Identify and correctly name at least 90% of your devices.

A mostly-labeled network is exponentially more manageable than a perfect one that never gets finished.

HOW TO: Access Your Device List

1. Open the Firewalla app

2. Tap 'Devices' in bottom navigation
3. Tap arrows icon (top right) for View Options
4. Use search bar to find devices

HOW TO: Identify Unknown Devices

STEP 1: Check Basic Information
1. Tap device
2. Review Manufacturer, Hostname, MAC

STEP 2: Analyze Network Activity
1. Check Network Flows
2. See what services device connects to

STEP 3: MAC Lookup
1. Visit MAC lookup website
2. Enter first 6 characters

STEP 4: Elimination Method
1. Disconnect suspects one at a time
2. Check if unknown goes offline

HOW TO: Rename Devices

1. Tap device
2. Tap name
3. Enter new name
4. Save

AUTOMATING DISCOVERY WITH NEW DEVICE QUARANTINE

Manually scanning for unknown devices works, but Firewalla offers a powerful automation layer: New Device Quarantine. When enabled, every new device that joins your network is automatically placed into a locked-down Quarantine Group with pre-defined rules that block internet access and local network communication.

This is operationally significant. Instead of discovering unknown devices after they've already been active on your network, quarantine contains them at the moment of arrival. You decide what gets access — not the device.

Why this matters: IoT devices often use MAC address randomization (iOS 14+, Android 10+), which causes them to appear as "new" devices each time the address changes. Quarantine catches these automatically, preventing devices from evading your rules by randomizing their identity.

How quarantine works in practice: New devices are placed in the Quarantine Group automatically. The group comes with two pre-defined rules: block internet access and block local network access. You can customize these rules, add your own, or treat the Quarantine Group like any other device group. When you've identified and trusted a device, release it from quarantine and assign it to the appropriate group.

Operator insight: Let unknown devices run in quarantine for a day or two while you observe their network flows. Devices that only connect to trusted servers (like Amazon AWS or Google Cloud) are more likely to be legitimate IoT devices.

Devices attempting to reach unusual destinations deserve scrutiny before release.

HOW TO: Enable New Device Quarantine

1. Main screen > scroll down > tap the "+" more button
2. Tap New Device Quarantine
3. Toggle ON
4. On Gold/Purple: choose which networks to apply quarantine to
5. Return to Devices — you'll see a new Quarantine Group

HOW TO: Release a Device from Quarantine

1. Tap Devices > Quarantine Group
2. Swipe left on the trusted device
3. Tap "Leave Group"
4. Assign the device to its proper group (Trusted, IoT, etc.)

Tip: If you use quarantine as your default posture, disable private/random MAC addresses on your own devices to prevent them from being repeatedly quarantined. Keep quarantine active for guest-facing networks where unknown devices are expected.

TROUBLESHOOTING: Devices Not Appearing

1. Check network connection
2. Wait for low-power devices
3. Check 'Show Past Devices'
4. Force reconnect

OPERATOR TAKEAWAYS

- You now have a repeatable process for identifying unknown devices

- You understand which techniques scale and which don't

  You've reduced uncertainty—the enemy of good security decisions

NEXT: We lock this knowledge in with naming conventions that won't collapse later.

CHAPTER 2 NAMING CONVENTIONS THAT SCALE

Good naming is infrastructure. Every rule, alert, report, and decision depends on it.

Poor names multiply friction. Good names compound clarity.

WHY NAMING MATTERS OPERATIONALLY

Names appear everywhere: device lists, alarms, rules, bandwidth reports, and flow analysis. If names are ambiguous, every interaction costs time and attention.

A good name should answer three questions at a glance:

- What is this?
- Who or where does it belong?
- Why does it exist on the network?

THE THREE LAWS OF DEVICE NAMING

Self-Documenting — The name should explain itself six months from now.

Consistent — Similar devices follow the same pattern without exception.

Scalable — The convention must work at 10 devices or 100.

PROVEN NAMING PATTERNS

Person–Device (personal endpoints)

- Sarah-iPhone
- John-Laptop

Location–Device (IoT and shared devices)

- LivingRoom-TV
- FrontDoor-Camera

Hybrid (recommended) - Personal devices:

- Person–Device
- IoT devices: Location–Device

Stay consistent *within categories*.

HOW TO: Implement Your Naming Convention

1. Choose pattern (Person-Device, Location-Device, or Hybrid)
2. Start with important devices
3. Rename: Tap device > Name > Edit > Save
4. Aim for 90% completion

IF YOU ONLY DO ONE THING

Pick a convention today and apply it relentlessly. Imperfect consistency beats perfect indecision.

OPERATOR TAKEAWAYS - Names are operational leverage:

- Consistency reduces alert fatigue and rule errors

- Future-you can understand present-you's decisions

NEXT: We stop managing devices individually and start managing them as systems.

CHAPTER 3 GROUPING STRATEGY

Groups let you think and act at the right level of abstraction.

Without groups, rules multiply. With groups, rules scale.

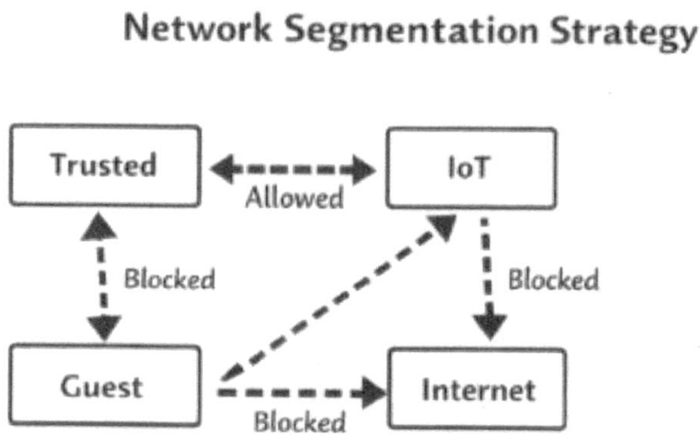

WHY GROUPS MATTER

Groups enable:

- One-to-many rule application
- Aggregate visibility
- Faster troubleshooting
- Reduced configuration drift

If you can't explain why a group exists, it shouldn't.

ESSENTIAL GROUPS

Trusted Group

- Personal computers and phones
- Full access with Minimal restrictions

IoT Group

- Smart home devices
- Internet allowed
- Local access restricted

Guest Group

- Visitor devices set to Internet only
- No local access

Kids (optional) — Child-used devices

- Content filtering
- Time controls

Start with 3–5 groups. Add only when there's a clear rule or visibility need.

GROUP MEMBERSHIP GUIDELINES

- Most devices belong in 1–2 groups
- Avoid excessive overlap
- Infrastructure devices (NAS, hubs, printers) deserve special treatment

---

HOW TO: Create a Device Group

1. Devices > Scroll to bottom > Create Group
2. Enter name (IoT, Trusted, Guest, Kids)
3. Optionally select devices
4. Save

HOW TO: Add Devices to a Group

METHOD 1 - From Group:

1. Tap group > Devices > Manage
2. Select devices
3. Save

METHOD 2 - From Device:

1. Tap device
2. Tap current Group
3. Select new group

IMPORTANT: Group Membership Rules

When device joins group:

- All device-level rules are DELETED
- Device adopts all group rules
- Device can only be in ONE group

IF YOU ONLY DO ONE THING:

- Create Trusted, IoT, and Guest groups and assign every device.

OPERATOR TAKEAWAYS:

- Groups are how rules scale safely
- Fewer, purposeful groups outperform many vague ones
- The foundation for security rules is now in place

NEXT: We enable baseline protections that work automatically.

CHAPTER 4 BUILT-IN PROTECTIONS

This chapter is intentionally practical. You should be toggling features while you read.

The goal here is not perfection. It's baseline protection that works quietly, automatically, and continuously.

WHY START WITH BUILT-IN PROTECTIONS

Firewalla's built-in protections deliver the highest security return for the lowest effort. They catch background noise, known bad actors, and common attack patterns before you ever need to think about custom rules.

These features work best when applied after devices are named and grouped. Now that your foundation is solid, we turn them on deliberately.

ACTIVE PROTECT — YOUR FIRST LINE OF DEFENSE

Active Protect is Firewalla's intrusion detection and prevention system. It monitors traffic in real time and blocks known malicious behavior.

What it blocks:

- Malware command-and-control traffic
- Exploit attempts and port scans
- Brute-force login attempts

- Known malicious destinations

Recommended setting: Default mode

Default mode balances protection and usability. Strict mode exists, but it increases false positives and operational overhead.

Enable it once. Leave it on.

WHAT TO EXPECT AFTER ENABLING

Seeing blocks is normal. The internet is noisy. Firewalla is filtering that noise for you.

If something legitimate breaks:

- Check blocked flows
- Identify the destination
- Create a narrow exception
- Avoid disabling protection globally to solve a local issue.

HOW TO: Enable Active Protect

1. Main screen > + button > Active Protect
2. Toggle ON
3. Choose Default mode (recommended)
4. Save

AD BLOCK — REDUCING TRACKING AND NOISE

Ad Block works at the network level using DNS filtering. It reduces advertising, tracking, and telemetry across *all* devices — including those that can't run browser extensions.

It is especially effective for:

- Smart TVs
- Streaming devices
- IoT hardware

Recommended approach:

- Enable network-wide, or Enable for IoT and Kids groups first.

Expect fewer ads, faster page loads, and less background chatter.

HOW TO: Enable Ad Block

1. Main screen > Ad Block
2. Toggle ON
3. Choose Default blocking level
4. Choose where to apply (All Devices or groups)
5. Optionally exclude specific devices
6. Save

FAMILY PROTECT — CONTENT FILTERING WITH INTENT

Family Protect filters content by category rather than by individual site. It should be applied selectively, not universally.

Typical use cases:

- Kids devices
- Guest devices (optional)
- Categories worth enabling:
- Adult content
- Gambling
- Illegal activity

Avoid over-filtering. Excessive blocking creates workarounds that undermine security.

BYPASS PREVENTION

Filtering only works if it can't be trivially bypassed.

For Kids groups, consider:

- VPN blocking
- Enforced safe search
- Blocking DNS-over-HTTPS
- Each layer closes a common escape hatch.

HOW TO: Enable Family Protect

1. Main screen > Family > Family Protect
2. Toggle ON
3. Choose mode (3rd-Party or Native)
4. Select devices/groups to protect
5. Customize categories (if Native mode)
6. Save

DNS STRATEGY — CLOSING THE BYPASS VECTOR

DNS is the address book of the internet. Every time a device visits a website, it first asks a DNS server for the address. This makes DNS the most common bypass vector for your protections — if a device can use its own DNS, it can circumvent your ad blocking, content filtering, and monitoring entirely.

Firewalla provides several layers of DNS control. Understanding them is essential for maintaining the protections you've enabled.

DNS Booster: This is Firewalla's DNS interception feature. When enabled, it forces all devices to use Firewalla as their DNS resolver, regardless of what DNS settings the device itself specifies. This prevents devices (or clever users) from bypassing your filtering by manually changing their DNS to 8.8.8.8 or another public resolver. DNS Booster is enabled by default and should remain on.

DNS over HTTPS (DoH): DoH encrypts DNS queries between Firewalla and your chosen DNS provider over HTTPS. This prevents your ISP from seeing which sites your network is requesting. Important: DoH operates between Firewalla and the DNS provider, not between your devices and Firewalla. Your devices still send unencrypted DNS to Firewalla over your LAN, which is how Firewalla maintains visibility and filtering.

Unbound: Unbound is a local recursive DNS resolver that runs directly on your Firewalla box. Instead of sending all your DNS queries to a single provider like Cloudflare or Google, Unbound resolves domains by querying authoritative nameservers directly. This prevents any single DNS provider from having a complete record of your browsing activity. For additional privacy, you can enable Unbound over VPN.

Custom DNS Rules: You can create local DNS entries via the app (Services > Custom DNS Rules). These act as a local address book — useful for pointing custom hostnames to local devices like a NAS or home server.

NTP Intercept: Many devices make Network Time Protocol requests to sync their clocks. Firewalla's NTP Intercept catches these requests and processes them locally using trusted NTP servers. This reduces exposure while ensuring devices keep accurate time.

BLOCKING DNS BYPASS ATTEMPTS

Even with DNS Booster active, some applications and devices attempt to bypass your DNS controls using DNS over HTTPS (DoH) at the application level — particularly browsers like Chrome and Firefox, and some smart TV platforms.

Firewalla includes a built-in Target List called "DoH Services" that contains known DoH endpoints. You can create a block rule using this target list to prevent any device from resolving DNS through encrypted channels that bypass Firewalla's filtering.

HOW TO: Block Client-Side DoH Bypass

1. Rules > Add Rule
2. Action: Block
3. Target: Target List > DoH Services
4. Device: All Devices (or specific groups like IoT, Kids)
5. Save

This ensures that all DNS traffic passes through Firewalla, where your ad blocking, content filtering, and monitoring rules can take effect.

HOW TO: Enable DNS over HTTPS

1. Main screen > tap Services
2. Tap DNS over HTTPS
3. Toggle ON
4. Choose your DoH provider (Cloudflare, Google, Quad9, etc.)
5. Tap Apply To > select networks or groups
6. Save

HOW TO: Enable NTP Intercept

1. Main screen > tap Services
2. Tap NTP Intercept
3. Toggle ON
4. Tap Apply To > select All Networks
5. Save

IF YOU ONLY DO ONE THING

Enable:

- Active Protect (Default mode)
- Ad Block (network-wide or IoT-only)

This alone eliminates a significant amount of ambient risk.

OPERATOR TAKEAWAYS

- Baseline protection should be automatic and boring
- Built-in features handle the majority of background threats
- Custom rules work *better* once these protections are active

NEXT: We begin isolating devices so compromise doesn't spread.

CHAPTER 5 ONE-WAY COMMUNICATION RULES

Most devices on your network do not need to talk to each other.

They need to talk **out** to the internet. And they need to respond **in** when you initiate a connection. That's it.

Anything beyond that increases risk without adding value.

THE PROBLEM WE'RE SOLVING

By default, networks are flat. Every device can attempt to communicate with every other device. This is convenient — and dangerous.

IoT devices are the weak link:

- Minimal security updates
- Weak authentication
- Cloud-first designs
- Little visibility when something goes wrong

If one device is compromised, a flat network lets that compromise spread.

THE CORE PRINCIPLE: ASYMMETRIC TRUST

One-way communication means:

- IoT devices may initiate connections to the internet
- Trusted devices may initiate connections to IoT devices

- IoT devices may **not** initiate connections to trusted devices

This preserves functionality while limiting blast radius.

You control your devices. They don't control you.

THE BASE RULE

Create a rule that blocks IoT devices from accessing local networks.

Action: Block
Target: All Local Networks
Apply to: IoT group
Direction: Outbound

This single rule does most of the work.

HOW TO: Create IoT-to-Local Block Rule

1. Rules > Add Rule
2. Action: Block
3. Target: All Local Networks
4. Device: IoT group
5. Direction: Outbound
6. Save

Understanding Direction:
• Outbound = IoT device initiating
• You can still access IoT devices

HOW TO: Create Exceptions

Option 1: Remove device from IoT group
Option 2: Create Allow rule for specific device

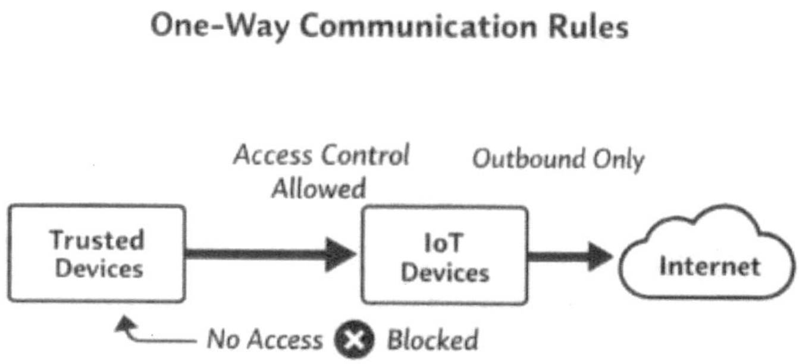

WHAT CONTINUES TO WORK

After this rule:

- Cameras still upload footage'
- Smart speakers still respond
- You can still view and control devices

Stateful firewalls allow replies to your requests. You are blocking *initiation*, not *responses*.

COMMON EXCEPTIONS (AND WHY THEY EXIST)

Some devices rely on local discovery or coordination.

Typical cases:

- Casting devices (Chromecast, AirPlay)
- Smart home hubs
- Printers

Treat these as infrastructure, not generic IcT.

Options:

- Place them in separate groups
- Create narrow allow rules
- Avoid weakening the IoT group globally
- Exceptions should be specific, documented, and rare.

TESTING YOUR RULES

Verify three things: 1. Devices still function normally

2. You can still access them

3. Blocked flows appear for IoT-to-local attempts

If functionality breaks, you blocked too much.

SCALING YOUR RULES WITH TARGET LISTS

As your rule set grows, managing individual domains and IP addresses becomes unwieldy. Target Lists solve this by letting you group related domains and IPs into named lists that can be referenced in any rule.

Think of target lists as building blocks for rules. Instead of creating ten separate block rules for ten ad-tracking domains, you create one target list containing all ten domains and one rule that references the list. When you need to add or remove entries, you update the list — not every rule that uses it.

BUILT-IN TARGET LISTS

Firewalla ships with several pre-built target lists that enhance your security without additional configuration:

DoH Services: Contains known DNS-over-HTTPS endpoints. Block this list to prevent devices from bypassing your DNS filtering (covered in Chapter 4).

Apple Private Relay: Contains Apple's iCloud Private Relay endpoints. Blocking this prevents iOS devices from routing traffic through Apple's relay, which would bypass your monitoring and filtering.

OISD (Open Internet Security Data): A curated list of risky and unwanted domains. Blocking this provides an additional layer of protection beyond Ad Block.

Newly Registered Domains (NRD): Domains registered in the last 30 days. Malware frequently uses newly registered domains for command-and-control communication. Blocking NRDs can stop compromised devices from phoning home. Note: this may occasionally block legitimate new services.

CREATING CUSTOM TARGET LISTS

Custom target lists are created through the Firewalla Web Interface (my.firewalla.com) or the Firewalla MSP portal. You can add domains, IP addresses, or IP ranges to any custom list.

Common use cases: Whitelisting specific IoT services (all domains your Ring cameras need), creating a custom block list for your household, grouping work-related domains for policy-based routing, or muting alarms from known-safe IP ranges.

HOW TO: Use a Built-In Target List in a Rule

1. Rules > Add Rule
2. Action: Block (or Allow)
3. Target: Target List > select the list (e.g., OISD, NRD, DoH Services)
4. Device: Choose device, group, or All Devices
5. Save

HOW TO: Create a Custom Target List

1. Open my.firewalla.com in a browser
2. Navigate to Target Lists
3. Click Create Target List
4. Name your list (e.g., "Ring Camera Domains" or "Work Services")
5. Add domains, IPs, or IP ranges
6. Save — the list is now available in the app for use in rules

Operator tip: When you create an IoT allow-only rule (block all internet, then allow specific targets), build a target list of the domains each device needs. Observe the device's network flows for a few days to identify its required destinations before locking it down.

IF YOU ONLY DO ONE THING

Block IoT devices from initiating connections to local networks.

This single change dramatically reduces risk.

OPERATOR TAKEAWAYS

- One-way rules preserve usability while limiting damage
- Initiation matters more than visibility
- This is the first true containment layer

NEXT: We prevent compromised devices from moving laterally at all.

CHAPTER 6 BLOCKING LATERAL MOVEMENT

One-way rules limit damage. Blocking lateral movement stops it entirely.

If a device is compromised, the most important question is not *what it can reach on the internet* — it's *what it can reach next.*

WHAT LATERAL MOVEMENT IS

Lateral movement is when an attacker pivots from one compromised device to others on the same network.

Typical progression:

- Compromise a weak IoT device
- Scan the local network
- Attempt access to other devices
- Escalate privileges and expand control

Flat networks make this easy. Structured networks make it expensive.

WHY IOT-TO-IOT MATTERS

Even "low value" devices become dangerous in aggregate.

Cameras record video. Speakers have microphones. Hubs control automation. Once an attacker gains multiple footholds, visibility and persistence increase dramatically.

Stopping lateral movement keeps compromises isolated.

THE DEFAULT-DENY MODEL

Within the IoT group, devices generally do not need to communicate with each other.

The safest default is:

- IoT devices may talk outward to the internet
- Trusted devices may initiate connections to IoT
- IoT devices may **not** initiate connections to *any other local device*, including other IoT

THE CORE RULE

Create a rule that blocks IoT devices from accessing other IoT devices.

Action: Block
Target: IoT group
Apply to: IoT group
Direction: Outbound

HOW TO: Create IoT-to-IoT Block Rule

1. Rules > Add Rule
2. Action: Block
3. Target: IoT group
4. Device: IoT group (same)

5. Direction: Outbound
6. Save

IMPORTANT: Layer 3 Limitations

• Firewalla operates at Layer 3 (IP)
• Same-LAN devices can communicate via Layer 2
• For true isolation: VLANs or separate networks

HOW TO: Monitor for Lateral Movement

1. IoT group > Network Flows > Blocked
2. Look for: repeated attempts, scans, admin ports

This prevents device-to-device spread.

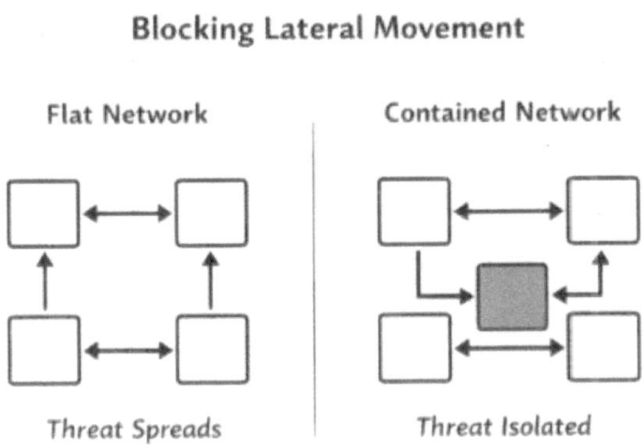

HANDLING LEGITIMATE COMMUNICATION

Some ecosystems require limited internal communication:

- Multi-room audio systems

- Hub-based automation platforms
- Certain lighting systems

Do not weaken the entire IoT group.

Instead:

- Isolate dependent devices into subgroups
- Allow only the minimum required protocols
- Keep allow rules narrow and documented

Infrastructure devices deserve special handling.

MONITORING FOR WARNING SIGNS

Blocked lateral attempts are useful signals.

Pay attention to:

- Repeated blocked connection attempts
- Sequential scans across multiple IPs
- Attempts against administrative ports

This behavior is rarely benign.

BEYOND LAYER 3: TRUE ISOLATION WITH VLANS

The rules covered above operate at Layer 3 (the IP level). They are effective for controlling traffic that routes through Firewalla, but devices on the same LAN segment can still communicate at Layer 2 (the data link level) without Firewalla seeing or blocking the traffic.

For true device isolation, you need network segmentation — physically or virtually separating devices into different networks. VLANs (Virtual Local Area Networks) accomplish this by dividing a single physical network into multiple isolated virtual networks, each with its own IP address range and its own set of rules.

What VLANs give you: Each VLAN is a separate broadcast domain. Devices on different VLANs cannot communicate at Layer 2. All inter-VLAN traffic must route through Firewalla, where your rules take full effect. This closes the Layer 2 gap that device groups alone cannot address.

Hardware requirements: VLANs require either a managed switch that supports 802.1Q VLAN tagging, or a VLAN-capable access point like the Firewalla AP7. If you only use Wi-Fi devices, the AP7 alone is sufficient — it can assign devices to VLANs based on which SSID or personal key they use to connect.

Practical VLAN strategy: A common home setup uses three VLANs: your main home network (default LAN), an IoT VLAN for smart devices, and a Guest VLAN for visitors. Each VLAN gets its own rules — for example, the IoT VLAN might block all access to other local networks, while the Guest VLAN blocks local access and enables content filtering.

VLAN limits: Firewalla Gold has no VLAN limit. Purple and Orange are limited to 5 VLANs. This is more than sufficient for most home networks.

HOW TO: Create a VLAN

1. Main screen > Network Manager
2. Tap Create Network > Local Network
3. Name the network (e.g., "IoT" or "Guest")
4. Set Type to VLAN
5. Set a VLAN ID (e.g., 66 for IoT, 99 for Guest)
6. Choose the LAN port
7. Use "Surprise Me" for IP settings, or set manually (convention: use the VLAN ID as the third octet, e.g., 192.168.66.x for VLAN 66)
8. Save

HOW TO: Isolate a VLAN from Other Networks

1. Rules > Add Rule
2. Action: Block
3. Target: All Local Networks
4. Device: Select the VLAN network (e.g., IoT VLAN)
5. Direction: Both
6. Save
7. Create additional Allow rules for any specific cross-VLAN access you need

Important: Device Groups and VLANs serve different purposes. Groups apply rules to a set of devices regardless of their network. VLANs physically isolate traffic at the network level. For maximum security, use both: segment devices into VLANs for Layer 2 isolation, then use groups within VLANs for more granular rule management.

VqLAN: MICROSEGMENTATION WITHOUT HARDWARE CHANGES

If VLANs feel like too much infrastructure, Firewalla offers a simpler alternative: VqLAN (Virtual Quarantine LAN). VqLAN provides Layer 2 isolation within your existing network — no managed switches, no IP address changes, no VLANs to configure.

VqLAN works by creating microsegments within a device group. When you enable VqLAN on a group, devices in that group can communicate with each other and access the internet, but they are blocked from reaching any device outside the group at Layer 2. This is true isolation, not just a firewall rule.

Requirements: VqLAN requires a Firewalla AP7 or Firewalla Orange. All devices in the VqLAN group must be connected to Firewalla Wi-Fi via the AP7 or Orange for microsegmentation to take effect.

Device Isolation: For even stronger control, enable Device Isolation within a VqLAN group. This prevents devices in the same group from communicating with each other — each device can only reach the internet and nothing else. This is ideal for IoT devices that have no legitimate reason to talk to any local device.

Allowed Devices: When VqLAN or Device Isolation is enabled, you can selectively permit specific devices to communicate using the Allowed Devices feature. For example, isolate your printer but allow your laptop and phone to print to it.

HOW TO: Enable VqLAN on a Group

1. Devices > select your group (e.g., IoT)
2. Scroll down to the VqLAN toggle
3. Toggle ON
4. Optionally enable Device Isolation for full per-device isolation

5. To allow specific devices through: tap Allowed Devices > select trusted devices

When to use VLANs vs. VqLAN: Use VLANs when you need full network-level segmentation, have managed switches, or need wired device isolation. Use VqLAN when your devices are primarily on Wi-Fi via the AP7, you want quick isolation without infrastructure changes, or you need device-level granularity within a group. The two can be combined for layered security.

IF YOU ONLY DO ONE THING

- _Block IoT devices from initiating connections to other IoT devices.

This closes the most common internal attack path.

OPERATOR TAKEAWAYS

- Containment only works if spread is impossible
- Lateral movement is how small problems become big ones
- Default-deny inside IoT dramatically improves resilience

NEXT: We use geographic controls to reduce external attack surface.

CHAPTER 7 REGION & COUNTRY BLOCKING

Not all traffic is equal.

If your network has no reason to communicate with certain regions of the world, allowing that traffic only increases exposure. Geographic filtering is a blunt tool—but used carefully, it meaningfully reduces noise and risk.

WHAT GEO-BLOCKING DOES (AND DOESN'T)

Region and country blocking limits connections based on the geographic location of IP addresses.

It **does**:

- Reduce automated scanning and background attacks
- Shrink the pool of external sources that can reach your network
- Lower alert noise so real issues stand out

It **does not**:

- Stop a determined attacker using VPNs or proxies
- Replace proper segmentation and containment
- Eliminate the need for monitoring

Think of geo-blocking as attack-surface reduction, not a silver bullet.

WHY THIS WORKS IN PRACTICE

Large portions of malicious traffic originate from a relatively small number of regions. Blocking those regions removes a disproportionate amount of hostile noise.

For most home networks, legitimate traffic is geographically narrow. If you live in North America, most services you use are hosted there or in Western Europe.

After enabling geo-blocking, review blocked flows after a few days before expanding your country list. Adjust incrementally. Anything outside that footprint deserves scrutiny.

A CONSERVATIVE STARTING POINT

Begin with a minimal block list:

- Countries you have no legitimate need to communicate with
- Regions known for high volumes of automated attacks

This avoids breaking services while still reducing exposure.

Avoid allow-listing entire continents early. Overblocking creates fragility.

APPLYING GEO-BLOCKING STRATEGICALLY

Recommended targets:

- IoT groups
- Guest devices

HOW TO: Enable Geo-Blocking

1. Rules > Add Rule
2. Action: Block
3. Target: Region > Select countries
4. Device: IoT group (recommended)
5. Direction: Outbound
6. Save

Strategy: Week 1 = 3-5 countries, Week 2 = 10-15

HOW TO: Create Exceptions

1. Check Blocked Flows
2. Create Allow rule for specific domain/IP

Be cautious applying geo-blocking to:

- Work devices
- VPN endpoints
- Services that rely on global CDNs

Different groups may require different geographic rules.

ADVANCED RULE OPTIONS

As your rule set matures, several advanced options become important for precision and maintainability:

Rule Direction: Rules can be set to Inbound, Outbound, or Both. Outbound blocks the device from initiating connections to the target. Inbound blocks traffic from the target reaching the device. For geo-blocking, Outbound is usually sufficient since you're preventing your devices from contacting those regions.

DNS Blocking: By default, internet blocking rules only block IP connections — DNS requests are still resolved, but traffic is blocked at the IP level. For stricter enforcement, enable DNS Blocking on a rule. This prevents the DNS query itself from resolving, ensuring no connection attempt is made at all.

Schedule-Based Rules: Rules can be scheduled to activate only during specific times. This is useful for parental controls (block gaming after 9 PM) or work-from-home policies (route work traffic through VPN during business hours).

Rule Notes: Firewalla includes a Notes field on rules, routes, Smart Queue rules, and custom DNS rules. Use this field to document why a rule exists and when it was created. This directly supports the operator principle of documenting decisions — future-you will thank present-you.

Target Categories: Firewalla provides system-managed target categories including Gaming, Social, Video, Adult, Gambling, and P2P. These categories are automatically populated by learning your network's traffic patterns. You can use them in rules just like domains or IPs, and you can view and edit the contents of each category.

HANDLING FALSE POSITIVES

When something breaks after enabling geo-blocking:

- Check blocked flows
- Identify the country being denied

- Determine whether the service is legitimate

Fix the problem narrowly:

- Create an exception for the specific service
- Adjust the country list incrementally

Avoid disabling geo-blocking entirely unless necessary.

TRAVEL AND REMOTE ACCESS CONSIDERATIONS

If you access your network remotely while traveling:

- Your own traffic may originate from blocked regions
- VPN connections may terminate in unexpected countries

Plan ahead. Temporarily relax geographic rules when traveling, then restore them afterward.

IF YOU ONLY DO ONE THING

- Block regions you have no legitimate reason to communicate with—starting conservatively.

Noise reduction improves visibility.

OPERATOR TAKEAWAYS

- Geo-blocking reduces exposure and alert fatigue
- It works best when applied selectively
- Conservative changes outperform aggressive ones

NEXT: We shift from configuration to ongoing operational awareness.

CHAPTER 8 OPERATIONAL AWARENESS & ONGOING MAINTENANCE

A well-run network is quiet.

Not silent — but predictable. Alerts mean something. Changes stand out. Most days, nothing requires action.

That calm is not accidental. It's the result of habits.

THE GOAL: BORING SECURITY

Good security does not demand constant attention. It reduces uncertainty until what remains is actionable.

If your Firewalla dashboard constantly demands intervention, something upstream is wrong — naming, grouping, or rule scope.

Operational awareness is about restoring signal-to-noise.

UNDERSTANDING NORMAL BEHAVIOR

Before you can spot problems, you must know what "normal" looks like.

Normal includes:

- Routine outbound connections
- Periodic update traffic
- Predictable usage patterns by time of day

- Spend time observing flows when nothing is wrong. This builds intuition that no rule can replace.

ALARMS THAT DESERVE ATTENTION

Not all alerts are equal.

Pay attention to:

- Repeated blocked attempts from a single device
- New devices appearing unexpectedly
- Lateral movement blocks
- Geographic blocks triggered internally

Ignore:

- One-off background scans
- Transient DNS noise

Your attention is finite. Spend it deliberately.

A SIMPLE MAINTENANCE RHYTHM

Weekly (5–10 minutes):

Review alarms

Scan for unknown devices

Monthly (20–30 minutes):

Review blocked flows

Confirm group membership still makes sense

Remove stale devices

Quarterly:

- Revisit exceptions
- Tighten rules that drifted
- Routine prevents surprises.

WHEN SOMETHING BREAKS

Breakage is feedback.

When a rule causes disruption:

- Identify what changed
- Check blocked flows
- Fix narrowly

Avoid rolling back protections globally.

Precision beats panic.

EVOLUTION, NOT PERFECTION

Your network will change.

New devices arrive. Old ones leave. Requirements shift. Security posture should evolve alongside reality.

Resist the urge to chase perfection. Stability and clarity matter more.

REMOTE ACCESS WITH FIREWALLA VPN SERVER

If you travel or need to access your home network remotely, Firewalla includes a built-in VPN server — no subscription required. This lets you securely connect to your home network from anywhere, accessing local devices and benefiting from your Firewalla's protections even on public Wi-Fi.

Firewalla supports both WireGuard (recommended for speed and simplicity) and OpenVPN protocols. You can create individual VPN profiles for each device, and each profile can be managed like a local device — with its own rules, monitoring, and access controls.

Security best practice: Create a VPN profile for each device that needs remote access. Then restrict VPN access: block all local networks for the VPN profile, and create specific Allow rules only for the resources that device needs.

HOW TO: Set Up WireGuard VPN Server

1. Main screen > VPN > VPN Server
2. Choose WireGuard
3. Toggle ON
4. Create a profile for each remote device
5. Share the profile with your device (QR code or config file)
6. Install WireGuard client on your device and import the profile
7. Optional: Create rules to restrict VPN access to specific local resources

DDNS: Firewalla automatically provides a Dynamic DNS address so your VPN server is reachable even if your ISP changes your IP address. It may take up to one hour for DDNS to propagate after initial setup.

BANDWIDTH MANAGEMENT WITH SMART QUEUE

If your household experiences buffering during video calls or laggy gaming while other devices are downloading, Smart Queue Management (SQM) can help. SQM reduces bufferbloat — the delay caused when network buffers overflow during heavy usage.

Smart Queue works by intelligently prioritizing traffic so that real-time applications (video calls, gaming) get the bandwidth they need, even when other devices are saturating the connection. You can also set rate limits per device or per group to prevent any single device from monopolizing bandwidth.

Operator note: Smart Queue is a router-mode-only feature. Enable it from the main screen under Services. Start with the default settings and adjust only if you notice issues.

HOW TO: Weekly Review (5-10 min)

1. Alarms > Scan patterns > Archive noise
2. Devices > Check for unknowns > Rename

HOW TO: Monthly Review (20-30 min)

1. Review Blocked Flows > Create exceptions
2. Verify Group Membership > Move misplaced devices
3. Remove Stale Devices

HOW TO: Troubleshoot When Something Breaks

STEP 1: Check Blocked Flows

- Tap device > Network Flows > Blocked
- Note what was blocked and why

STEP 2: Identify the Rule
• Go to Rules
• Find the blocking rule

STEP 3: Create Exception
• Rules > Add Rule
• Action: Allow
• Target: Specific blocked item
• Save

IF YOU ONLY DO ONE THING

Establish a regular review habit.

Most problems are obvious *before* they become incidents — if you're looking.

OPERATOR TAKEAWAYS

- Quiet networks are healthy networks
- Attention should be spent intentionally
- Habits outperform complex configurations

NEXT: We consolidate everything into a sustainable operating model.

CHAPTER 9 THE SUSTAINABLE OPERATOR MODEL

The difference between a secure network and a fragile one is not how many rules exist. It is whether the system can be sustained without constant effort.

A sustainable network is one you can run calmly, confidently, and consistently — even as devices change and time passes.

FROM CONFIGURATION TO OPERATION

Setup is finite. Operation is ongoing.

Most failures happen when networks remain stuck in a setup mindset: reacting to individual problems instead of maintaining a system.

The operator model replaces reaction with posture.

THE FOUR PILLARS

A sustainable Firewalla deployment rests on four pillars:

1. **Clarity** — You know what is on your network and why
2. **Containment** — Compromises cannot spread
3. **Reduction** — Exposure is limited intentionally
4. **Awareness** — Attention is spent deliberately

Remove any pillar and the system degrades.

The Sustainable Operator Model

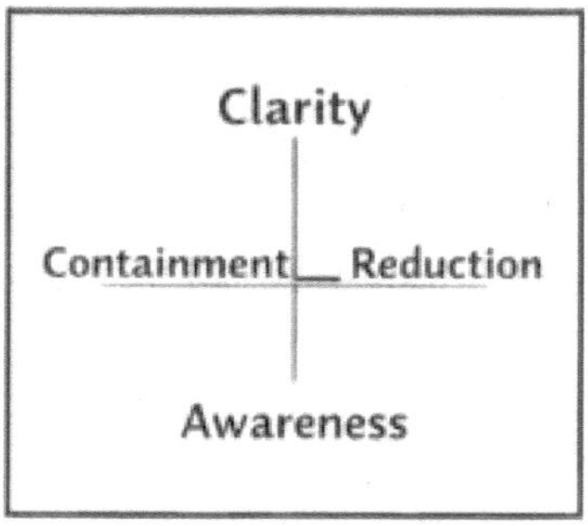

WHY THIS MODEL HOLDS UP OVER TIME

- Devices will change.
- Interfaces will change.
- Threats will change.

The model does not depend on specifics. It depends on judgment.

That is why it scales from a handful of devices to dozens without collapsing.

WHAT TO DO WHEN YOU FEEL DRIFT

Drift is normal.

When the network starts to feel noisy or brittle:

- Revisit naming and grouping
- Review exceptions
- Remove rules that no longer serve a purpose

Return to fundamentals before adding complexity.

RESISTING SECURITY THEATER

More rules do not equal more security.

Security theater looks busy but produces little resilience. The operator model values effectiveness over appearance.

If a rule does not reduce risk or improve clarity, it does not belong.

THE LONG VIEW

A well-run network fades into the background.

It supports daily life quietly. It does not demand constant attention. When something does require action, it is clear and understandable.

That is the goal.

THE ZERO TRUST FRAMEWORK — AN ADVANCED TRACK

For operators who have mastered the fundamentals in this book, Firewalla supports a formal Zero Trust architecture that takes the Four Pillars model to its logical conclusion: trust nothing by default, verify everything explicitly, and grant only the minimum access necessary.

Zero Trust rests on three principles that map directly to the concepts you've already learned:

Microsegmentation (Containment): Divide your network so that each device or group of devices is isolated from the rest. You've already done this with groups and one-way rules. VLANs and VqLAN (covered in Chapter 6) take this to Layer 2, where isolation is complete.

Least Privilege Access (Reduction): Grant each device only the minimum permissions it needs to function. For IoT devices, this means identifying the exact domains and services they require (using network flows), creating a Target List of those destinations, blocking all other internet access, and allowing only the listed targets.

Continuous Monitoring (Awareness): Continuously verify the security posture of all devices. Your weekly and monthly review habits from Chapter 8 serve this purpose. Firewalla's alarms, network flows, and blocked flow reports provide the data you need.

DEVICE ACTIVE PROTECT (DAP)

Device Active Protect is Firewalla's automated least-privilege enforcement for IoT devices. When enabled, DAP analyzes eligible devices' traffic patterns and automatically restricts each device to only the trusted sites it needs to function. Everything else is blocked.

DAP works best on simple, purpose-specific IoT devices like thermostats, sensors, and cameras that consistently communicate with a small set of cloud services. More complex devices like laptops and phones are not eligible.

HOW TO: Enable Device Active Protect

1. Main screen > Protect
2. Enable Device Active Protect
3. If prompted, enable NTP Intercept for all networks
4. Eligible devices will be automatically protected

Requirements: App version 1.66+ and Box firmware 1.981+. Available on all Firewalla Gold series boxes in Router or Bridge mode.

BUILDING YOUR ZERO TRUST NETWORK

You do not need to implement Zero Trust all at once. The work you've done in this book is already the foundation. Here is a progression path:

1. Start with the groups, rules, and protections from Chapters 1–8 (you are here)

2. Enable VqLAN or VLANs for true device isolation (Chapter 6)

3. Enable Device Active Protect for eligible IoT devices

4. Build Target Lists for devices that need manually configured least-privilege access

5. Enable New Device Quarantine to contain unknown devices by default (Chapter 1)

6. Set up VPN server for secure remote access (Chapter 8)

Each step builds on the last. The system remains sustainable because each layer is designed to operate with minimal ongoing attention — which is exactly what this book has been teaching from the beginning.

IF YOU ONLY DO ONE THING

Operate your network like a system, not a project.

Sustainability is the real measure of success.

OPERATOR TAKEAWAYS

- Sustainable security prioritizes clarity over complexity
- Good posture beats constant tuning
- Calm operation is not accidental — it is designed
- You now know how to run your network — not just configure it.

If the system fades into the background and only asks for your attention when it matters, it's working.

That calm is designed.

END OF OPERATIONAL EDITION

If you've read through the entire *Personal Security in a Digital World* series, you've built something many people never do—a layered approach to protecting your digital life.

Let's consider what you've accomplished if you've read the entire series—and more importantly, implemented what you've learned.

A password manager, hardware security keys, and network security each solve different problems—but together they create *multiple layers of protection.*

Bitwarden helps you create, store, and manage strong, unique passwords for every account.

YubiKey adds a powerful second layer of authentication, helping protect email, financial accounts, and other important services from phishing and account takeovers.

Network Security Essentials introduces the principles of securing your home network using Firewalla and explains the reasoning behind effective security decisions.

Firewalla: After Setup shows how to put those principles into practice over time by using intrusion detection, network segmentation, traffic visibility, and thoughtful network management to keep your security effective as your network grows.

Each layer addresses a different part of your digital life.

Together, they create a stronger, more resilient security strategy than any one product can provide alone.

Your Security Foundation

- **Protect your identity.** *(Bitwarden Essentials)*
- **Protect your accounts.** *(Guard Your Money)*
- **Protect your network.** *(Network Security Essentials)*
- **Operate your network securely.** *(Firewalla: After Setup)*

We hope you enjoyed our books, and we wish you well on your journey, wherever it may lead you. Stay safe.

- **Prairie Light Press**

Other Books by Prairie Light Press

The Awareness Series

The Awareness Series explores the nature of thought, attention, and the sense of self through direct observation—not theory.

Written in clear, simple language, these books are designed to be **experienced, not analyzed**, helping readers notice awareness as it unfolds in real time.

Awareness *(Start Here)*

Noticing the Voice Inside Your Head

The starting point — recognizing the voice and stepping out of automatic identification with it.

Still Noticing

The continuation — living with awareness over time, including the parts that are often left out.

These books can be read in any order, but they follow a natural progression:

Awareness → Still Noticing

Explore the Complete Prairie Light Press Catalog

Visit Prairie Light Press
www.prairielightpress.com

Browse on Amazon
Amazon Author Page

Our Security Books by Prairie Light Press

All titles are part of the **Personal Security in a Digital World** *series.*

Bitwarden Essentials ***(Recommended Next)***

A practical guide to securing and managing your passwords using Bitwarden, with step-by-step instructions for setup, daily use, and account recovery.

Guard Your Money

A clear, practical guide to protecting financial accounts using YubiKey hardware security, covering phishing, identity theft, and modern account takeover threats.

Network Security Essentials

A plain-language guide to securing a home network using Firewalla Gold Plus, focused on understanding security choices rather than just technical settings.

Firewalla: After Setup

Firewalla: After Setup focuses on the decisions that matter once your network is running—how to structure devices, reduce noise, contain risk, and keep security effective over time without constant tweaking.

For further reading check out

Prairie Light Dispatch

Short observations on awareness, daily life, and the systems we build around them.

Quiet notes from the prairie.

Visit:
lgphil7.substack.com

Reader Preview

Awareness
Noticing the Voice Inside Your Head

The following pages are an exclusive preview of **Awareness**: *Noticing the Voice Inside Your Head*

Exclusive Preview from Prairie Light Press

Awareness

Noticing the Voice Inside Your Head

PRAIRIE LIGHT PRESS
ESTD. 2025

Helping readers move from the unknown to the known

Contents of the Complete Book
Page numbers refer to the full edition.

Table of Contents

Published by Prairie Light Press through IngramSpark

Awareness: Noticing the Voice Inside Your Head

Publication Identifiers:

ISBN (Hardback): 9798996980208
ISBN (Paperback): 9798996980277

This book is for informational purposes only and is not intended as medical, psychological, or professional advice.

Published by

Prairie Light Press

www.prairielightpress.com

Introduction

I didn't plan to write this book.

It grew out of something quieter and more personal — a gradual shift I noticed while working through anxiety, stress, and a long-standing habit of being at war with my own mind. I didn't know how to describe it at first. I just knew that something fundamental had changed.

A therapist I was working with gave it a name: **awareness**.

At the time, that word meant almost nothing to me. It sounded vague. Abstract. Like something people talked about but never quite explained. But she was right — not because awareness fixed anything, but because it revealed something that had been there all along.

This book is not about improving yourself.

It's not about controlling your thoughts, mastering your emotions, or becoming a calmer, better version of who you are. I tried all of that. It only made things louder.

What changed wasn't my thinking.

It was my relationship to it.

Over time, I began to notice something simple but profound: thoughts didn't need obedience. Emotions didn't need management. Urges didn't need suppression. And most importantly, the mind didn't quiet down because I fought it — it quieted down because I stopped feeding it my attention.

That single realization altered everything.

This book is a record of that recognition — how awareness became obvious, how interference became visible, and how peace emerged not through effort, but through subtraction.

You won't find techniques here. No routines. No practices to perfect. Awareness isn't something you do. It's something you notice — usually right after you stop trying.

My hope is that by sharing this clearly and honestly, it helps you recognize the same thing in your own experience. Not as a belief. Not as a philosophy. But as something immediately verifiable.

If you're tired of managing yourself, fixing yourself, or arguing with your own mind, this book is for you.

Nothing here will ask you to become someone else.

It will simply invite you to notice what's already present — and to stop interrupting it.

— L. Phillips
February 2026

Awareness

Chapter 1: Something Changed (And I'm Not Sure When)

I can't pinpoint the exact moment.

That's the strange part. You'd think something this significant would have a timestamp -- a before and after, a clear line in the sand. But it wasn't like that. It was more like waking up one morning and realizing the background noise that had been running for years had gotten quieter.

Not silent. Just... quieter.

For most of my adult life, I'd been at war. Not with other people. Not with circumstances. With myself. With my thoughts. With this constant internal narrator that had opinions about everything I did, critiques ready before I even finished a sentence, judgments lined up like ammunition.

You know that voice. The one that says you should have done it differently. That you're not doing enough. That other people have it figured out and you're just faking it. That you need to manage yourself better, control yourself more, fix whatever's wrong before anyone notices.

I didn't realize how exhausting it was until it started to ease.

The Trading Floor Inside My Head

I trade markets for a living. Not the romantic version you see in movies—I'm not on a trading floor yelling into phones. I'm at home, watching screens, making decisions that sometimes work out and sometimes don't. It's a good metaphor for what was happening inside my head, though.

Markets are chaotic when you don't understand them. Everything feels urgent. Every move feels like it means something about you. Win, and you're smart. Lose, and you're an idiot. The volatility isn't just in the charts—it's in your nervous system.

When I first started trading, I was a mess. Every losing trade felt personal. Every missed opportunity gnawed at me. I'd lie awake replaying decisions, convincing myself I should have known better, beating myself up for not being perfect.

Sound familiar?

But over time, something shifted in how I approached the markets. I stopped trying to control them. I stopped making every outcome mean something about me. I learned to map the territory—to understand the patterns, the rules, the way things actually work rather than how I thought they should work.

Unknown became Known. Known became Manageable.

Not easy. Just manageable.

And here's what I didn't expect: that same process started happening in the rest of my life. Without me trying.

The Shift No One Tells You About

I started noticing something strange.

Thoughts would show up—the same critical, anxious, should-be-doing-better thoughts—and I'd just... see them. Not fight them. Not agree with them. Not try to fix them. Just see them.

Like watching a car drive by. It's there, then it's not. No need to chase it. No need to get in.

Urges would arise—to check my phone, to eat when I wasn't hungry, to react when someone said something that used to push my buttons—and they'd just... pass. Not because I white-knuckled through them. Because I noticed them, and then they faded on their own.

I wasn't managing myself better. I wasn't controlling myself more. I was just getting out of my own way.

And life started to feel different.

At Peace, Not War

That's the phrase that keeps coming back to me: At peace, not war.

I didn't win some battle against my mind. I didn't conquer my emotions. I didn't achieve some zen state where nothing bothers me. That's not what this is.

I just stopped fighting.

And when you stop fighting yourself, something remarkable happens.

- The system starts to regulate itself.
- Anxiety doesn't spiral because you're no longer feeding it with resistance.
- Self-criticism doesn't land because you're not taking it so seriously.
- Outcomes—wins, losses, whatever life throws at you—stop defining you because you're not building your worth on them.

It sounds too simple. I know. I would have dismissed this as naive or oversimplified a few years ago.

But here's what I've learned: the most profound shifts are often the simplest. Not easy—simple. There's a difference.

The Thing Everyone Already Has (But Doesn't Notice)

This book isn't about giving you something new.

You don't need a new technique. You don't need a practice. You don't need to become someone different or achieve some higher state.

You already have what this book is pointing toward. You've had it your whole life.

It's called awareness.

Not the buzzy, wellness-industry version of awareness. Not mindfulness as another item on your to-do list. Not some spiritual concept that requires years of meditation to access.

Just... awareness. The thing that's reading these words right now. The thing that noticed you were tense just then (or didn't). The thing that knows when you're distracted, anxious, angry, calm, whatever.

That. You already have that.

The only "work" involved—if you can even call it work—is to notice it. To notice that something is noticing. To catch yourself in the act of being aware.

And then to stop interrupting it.

What This Book Is (And Isn't)

This isn't a "do these steps and you'll be happy" book. I'm not going to give you a seven-day challenge or a morning routine or a journaling prompt.

This is more like... pointing at something. Something you already know but maybe haven't fully recognized yet.

I'm going to share what I discovered—not because I'm special or because I cracked some code, but because I stumbled into something that changed everything. And it might be useful to you. Or to someone you care about. Or to your clients, if you're a therapist (which is partly why I'm writing this).

You'll notice I don't talk down to you. I don't shame you for being stressed or anxious or self-critical. Why would I? You're not broken. You're not doing it wrong.

You're just caught up in a pattern that most humans get caught up in:

- believing every thought
- obeying every urge
- letting every outcome define you

And there's a way out that doesn't involve more effort.

It involves less.

A Warning (The Good Kind)

If you're looking for strategies, techniques, and action steps, this might frustrate you. This book is more about subtraction than addition. More about recognizing than achieving. More about allowing than controlling.

But if you're tired—tired of being at war with yourself, tired of managing your mind, tired of feeling like you're never quite enough—then keep reading.

Because what I'm about to share isn't about becoming better.

It's about stopping the interference. And letting what's already there come forward.

It's about noticing. And then letting that be enough.

Let's begin.

Next: Chapter 2 - "You Didn't Do Awareness. You Stopped Interrupting It."

Chapter 2: You Didn't Do Awareness. You Stopped Interrupting It.

Here's what nobody tells you about awareness:

You can't *do* it.

You can't achieve it, acquire it, or master it. You can't earn it through meditation retreats or morning routines or breathing exercises. You can't get better at it like you get better at piano or coding or trading.

Because you already have it. You've always had it.

The only thing you can do—and this is the whole game—is **stop interrupting it**.

That's it. That's the entire practice, if you can even call it a practice.

The Interference Pattern

For most of my life, I didn't know I was interrupting anything. I thought I was just... living. Reacting. Managing. Trying to stay on top of things.

But here's what was actually happening:

A thought would show up.

- And immediately—without even realizing it—I'd treat it like a command. Like something I had to obey, analyze, argue with, or fix.

- "I'm not good enough at this."
  Okay, better work harder. Prove that wrong. Fix it.

- "I should have done that differently."
  Time to replay it seventeen times, figure out what I did wrong, make sure I never do it again.

- "This might not work out."
  Better prepare for disaster. Game out every worst-case scenario. Protect yourself.

Every. Single. Thought.

I was in constant conversation with my own thinking, treating every mental visitor like it deserved an immediate response. Like thoughts were truths that required action.

An urge would arise.

And I'd either obey it immediately or fight it with everything I had.

The urge to check my phone during important work? Either surrender instantly or white-knuckle through it, gripping the desk like I'm resisting the pull of gravity.

The urge to eat when I wasn't hungry? Either give in and feel guilty, or force myself not to and feel deprived.

The urge to react when someone said something irritating? Either snap back or swallow it down, building resentment.

There was no middle ground. No space. Just reaction or resistance.

An outcome would land.

And I'd let it define me. Completely.

Good trade? I'm smart. I've got this figured out.
Bad trade? I'm an idiot. I'll never get this right.

Successful meeting? I'm capable. People respect me.
Awkward interaction? I'm socially broken. Everyone saw it.

My entire sense of worth swung with every result like a pendulum. And I didn't even notice I was doing it.

This is what I mean by interrupting awareness.

Awareness was there the whole time, noticing all of it—the thoughts, the urges, the outcomes. But I kept stepping in, managing, controlling, reacting. I kept getting in the way.

The Moment Things Shifted

I can't tell you exactly when it changed. But I can tell you what changed.

I started noticing the thoughts... without needing to do anything about them.

Not because I decided to practice some technique. Not because I read a book that told me how. It just... started happening.

A critical thought would show up—"You're screwing this up"—and instead of immediately engaging with it, I'd just... see it. Like watching a bird fly past the window.

Oh. There's that thought again.

And then it would pass. Not because I pushed it away. Not because I reasoned with it. Just because that's what thoughts do when *you don't feed them with your attention.*

Urges started doing the same thing.

That pull to check my phone during focused work? I'd notice it. Feel it. And then... it would fade. Not because I fought it. Because I didn't interrupt the natural cycle.

Urges arise. If you don't obey them and you don't resist them, they just... pass. That's what they do. They're designed to pass.

Outcomes stopped gripping me the same way.

A losing trade would happen, and instead of spiraling into self-criticism, I'd just notice: "Okay. That didn't work out. Next."

A winning trade would happen, and instead of inflating my ego, I'd just notice: "That worked. Good."

The outcome was just... information. Not a verdict on my worth.

What Actually Happens When You Stop Interfering

Once I stopped interrupting awareness—once I stopped constantly managing, analyzing, and reacting—something remarkable happened.

The system started regulating itself.

I don't mean that metaphorically. I mean literally, physically, mentally—things started to settle on their own.

Urges would arise and pass. I didn't have to fight them or give in to them. They just moved through.

Thoughts didn't need my obedience. They could show up, be noticed, and dissolve without me engaging in a twenty-minute internal debate.

Outcomes stopped defining me. I could still care about results—I absolutely did—but they didn't dictate my sense of self anymore.

And here's the wildest part: I started staying with experience instead of managing it.

Before, if I felt anxious, I'd immediately try to fix it. Breathe differently. Distract myself. Rationalize it away. Do something.

Now? I'd just... feel it. Notice it. Let it be there.

"Oh. Anxiety is here right now."

And you know what happens when you do that? When you just let experience be what it is without trying to manage it?

It settles. Naturally. On its own.

Not because you controlled it. Because you stopped interfering with it.

The Line That Says Everything

There was a moment—I don't even remember the exact context—when I realized how different things felt.

Someone asked me how I was doing with all of this. The trading. The stress. The mental noise.

And I said, almost without thinking:

"Just staying aware of whatever is going on with me... I like it."

That was it. That was the whole thing, distilled into one sentence.

Not "I've mastered my mind." Not "I've eliminated anxiety." Not "I've achieved some higher state."

Just: I'm staying aware of what's happening. And I like it.

That's not philosophy. That's not a concept I read about and decided to believe.

That's **embodied change**.

And it's the most stable kind—because it's not held together by effort. I'm not white-knuckling my way through some practice. I'm not forcing myself to be present. I'm just... noticing. And that's enough.

The Quiet Thesis

If I had to capture what this book is really about—what this whole shift comes down to—it would be this:

- **I didn't understand awareness. I just stopped getting in its way—and my life started to feel different.**
- I didn't learn some new skill. I didn't achieve some enlightened state. I didn't become a different person.
- I just stopped interrupting what was already there.
- And when I did, things quieted down. Not because I silenced them. Because I stopped fighting them.

You're not chasing anything anymore when this happens. You're not trying to get somewhere or become someone.

You're just... noticing.

And the fact that you're content to leave it there—that you don't need to analyze it, protect it, or turn it into something bigger—that's the strongest sign this isn't just another self-improvement phase.

You don't need to protect it. You don't need to prove it. You don't need to understand it.

You just keep doing what you're already doing:

Noticing... and letting that be enough.

What This Looks Like in Real Life

Let me give you a concrete example from trading, because it's where I saw this most clearly.

Before: I'd place a trade, and then I'd *watch* it. Not just monitor it—obsess over it. Every tick up felt like validation. Every tick down felt like impending disaster.

My stomach would tighten. My jaw would clench. I'd refresh the screen constantly, like I could control the outcome through sheer attention.

And if it went against me? The spiral would begin. "I knew this was a bad idea. Why did I do this? I should have waited. I'm terrible at this."

All of that was me interrupting awareness.

After: I'd place a trade, and I'd... notice I placed a trade. I'd notice the urge to obsess. I'd notice the tension in my body. I'd notice the thoughts trying to predict the future.

And I'd just let all of that be there without engaging with it.

The trade would play out however it played out. Sometimes it worked. Sometimes it didn't. But my nervous system wasn't hijacked by it anymore.

Because I wasn't interrupting. I was just aware.

The same thing happens in conversations now. In relationships. In daily stress.

Something irritating happens. I notice the irritation. I notice the urge to react. And then... it passes. Or it doesn't, and I respond—but from a different place. Not from reactive defensiveness. From awareness.

This isn't magic. This isn't some superhuman level of control.

It's just what happens when you stop getting in your own way.

You Already Know How to Do This

Here's the thing I want you to understand:

- You're not learning something new right now. You're recognizing something you've been doing your whole life.
- Every time you've ever had a moment of clarity—when you stepped back from a situation and just *saw* it without being caught up in it—that was awareness.
- Every time you've felt something difficult and just let it be there without trying to fix it—that was you not interrupting.
- Every time a thought passed through your mind and you didn't chase it or fight it—that was the system regulating itself.

You already know how to do this.

The only difference now is that you're starting to notice it. To recognize it. To trust it.

And once you do, you realize you don't need another technique, another practice, another program.

You just need to stop interrupting what's already working.

Next: Chapter 3 - "Unknown, Known, Manageable"

Chapter 3: Unknown, Known, Manageable

I have a way I operate.

It's not something I learned from a book or picked up in a seminar. It's just... how I'm wired. How I approach anything new, anything confusing, anything that doesn't make sense yet.

I call it: **Unknown, Known, Manageable.**

And once I recognized this pattern in myself, I realized it's exactly what happened with awareness. The same framework. The same progression. The same willingness to *pay an upfront cost to minimize pain downstream.*

Let me explain.

I'm a Systems Guy

Here's the thing about me: I like systems. I like understanding how things work. Not because I'm naturally brilliant or gifted—I'm not. But because I've learned that once I understand the system, everything becomes easier.

The problem is, when you first encounter any new system—software, trading platforms, writing tools, relationships, awareness—it's chaos. Pure chaos.

You don't know the rules. You don't know what buttons to push. You don't know what works and what doesn't. Everything feels hard because you're operating in the dark.

This is the **Unknown** phase.

And it's miserable. Frustrating. Confusing. Everything takes ten times longer than it should. You make mistakes you don't even know you're making. You feel stupid because you can't figure out what should be simple.

Most people quit here. They decide it's too hard, too complicated, not worth it. They move on to something else that feels easier.

But I don't quit in the Unknown phase. Not because I'm tough or disciplined. Because I know what comes next.

The Upfront Cost

Here's my operating principle:

I pay an upfront cost to minimize pain downstream.

In the Unknown phase, I don't try to get good at the thing yet. I don't try to be efficient. I don't try to achieve results.

I just... **map the territory.**

I poke. I prod. I click. I push. I pull. I try things that don't work just to see *why* they don't work. I make mistakes on purpose to understand the boundaries.

It looks chaotic from the outside. It looks inefficient. And in the short term, it absolutely is.

But what I'm actually doing is building a mental model. I'm learning the system. Not from a manual. Not from instructions. From direct experience.

This is the upfront cost. The confusion. The frustration. The time spent seemingly getting nowhere.

But I've learned something crucial: **This cost is temporary. And paying it now saves me exponentially more pain later.**

Because once I've mapped the territory, something shifts.

The **Unknown** becomes **Known**.

And once it's Known, it becomes **Manageable**.

From Chaos to Clarity

Let me give you a concrete example.

When I first started trading, I had no idea what I was doing. The platform was confusing. The terminology was foreign. The market movements felt random and terrifying.

Every trade felt like a gamble. Every loss felt personal. Every decision took forever because I didn't know what mattered and what didn't.

Unknown.

But I didn't quit. I paid the upfront cost.

I spent hours just watching. Not trading—just observing. Noticing patterns. Testing small positions. Making mistakes and noting what happened. Mapping the system.

I accepted that I would be confused. I accepted that I would lose money while I learned. I accepted the upfront cost because I knew it was buying me something valuable: understanding.

And eventually, something clicked.

The chaos started to organize itself. The patterns became visible. The rules became clear. Not because someone explained them to me, but because I'd mapped them through direct experience.

Known.

And once it was Known, it became **Manageable**.

Not easy. But manageable.

I could make decisions faster because I understood the system. I could see what mattered and what was just noise. I could trade without my nervous system being hijacked by every tick because I knew what was normal and what wasn't.

The same framework applies to literally everything I've ever learned:

- New software? Unknown → map it → Known → Manageable
- Writing and formatting? Unknown → map it → Known → Manageable
- Relationships? Unknown → map it → Known → Manageable

And yes, **awareness**.

How This Applies to Awareness

When I first started noticing this whole awareness thing, it was... confusing.

Unknown.

I didn't know what I was doing. I didn't know what "noticing" even meant half the time. I didn't know if I was doing it right. I didn't know why it sometimes felt clear and sometimes felt like I was just making it up.

It was chaotic. Frustrating. Like learning a new language without a dictionary.

Most people would quit here. They'd decide awareness is too vague, too mystical, too hard to pin down. They'd go back to the familiar pattern of managing thoughts, controlling urges, letting outcomes define them.

But I didn't quit. Because I recognized the phase I was in.

This is the Unknown phase. This confusion is the upfront cost.

So I did what I always do: I started mapping the territory.

I didn't try to "be aware" perfectly. I didn't try to achieve some enlightened state. I just... paid attention. Noticed what happened when I noticed. Observed the patterns.

I noticed:

- Sometimes a thought would show up and I'd get caught in it
- Sometimes a thought would show up and I'd just see it
- Sometimes I'd react to an urge, sometimes I wouldn't
- Sometimes outcomes would grip me, sometimes they wouldn't

I wasn't trying to control any of it. I was just mapping.

- "Oh, when I engage with that thought, this happens."
- "Oh, when I just let it be there, that happens."
- "Oh, when I stop trying to manage the urge, it fades on its own."

I was paying the upfront cost. The confusion. The uncertainty. The not-knowing-if-I'm-doing-it-right.

But I was learning the system.

And gradually, the **Unknown** became **Known**.

When It Becomes Known

You know that moment when you're learning something new and suddenly it just... clicks?

Not because someone explained it better. Not because you tried harder. Just because you'd been mapping the territory long enough that the patterns became visible.

That's what happened with awareness.

The chaos started to organize itself. The confusion cleared. I started to recognize what was happening:

- "Oh, I'm noticing a thought."
- "Oh, I'm interrupting awareness by engaging with it."
- "Oh, when I don't interrupt, it settles on its own."

It wasn't mystical. It wasn't magical. It was just... **Known**.

I'd mapped the territory through direct experience. I understood the system. Not intellectually—experientially.

And once it was Known, it became **Manageable**.

Again, not easy. But manageable.

I could recognize when I was interrupting and when I wasn't. I could see the difference between being caught in thought and noticing thought. I could feel when I was fighting experience versus staying with it.

And because I'd paid the upfront cost—the confusion, the uncertainty, the not-knowing—I didn't have to keep paying it over and over.

I'd learned the system. And now I could work with it instead of against it.

The Downstream Savings

This is why I pay upfront costs.

Because once something moves from Unknown to Known to Manageable, the ongoing effort drops dramatically.

Before I mapped awareness:

- Every moment of anxiety required active management
- Every critical thought demanded a response
- Every outcome felt like a verdict on my worth
- The mental energy required was **constant and exhausting**

After I mapped awareness:

- Anxiety could be there without needing to be fixed
- Thoughts could show up without needing to be engaged
- Outcomes could land without defining me
- The mental energy required dropped to almost nothing

That's the downstream savings.

I paid the confusion upfront. I accepted the Unknown phase. I did the work of mapping.

And now? I don't have to keep paying that cost. The system runs itself.

That's the trade. And it's worth it every single time.

You're in the Unknown Phase Right Now

If you're reading this and thinking, "I don't really get what he's talking about with this awareness thing"—good.

You're in the Unknown phase.

That's not a problem. That's not failure. That's just where you are in the progression.

The question is: Are you willing to pay the upfront cost?

Are you willing to sit with the confusion? To notice without knowing exactly what you're doing? To map the territory through direct experience instead of demanding clear instructions?

Because here's what I can tell you from the other side:

The Unknown doesn't last forever. If you stay with it—if you keep noticing, keep paying attention, keep mapping—it becomes Known.

And once it's Known, it becomes Manageable.

Not easy. But manageable.

And the downstream savings? Massive.

How to Map Awareness (Without Making It Complicated)

You don't need a formal practice. You don't need a meditation cushion. You don't need to set aside special time.

You just need to start noticing. That's the mapping.

- **Notice when you're caught in thought** versus when you're seeing thought.
- **Notice when you're fighting an urge** versus when you're just aware of it.
- **Notice when an outcome defines you** versus when it's just information.

You won't always know what's happening. That's fine. You're in the Unknown phase. The point isn't to get it right. The point is to map.

Pay attention. Observe the patterns. See what happens when you engage versus when you don't. Notice what changes when you stop interrupting.

That's it. That's the upfront cost.

It might feel confusing at first. It might feel like you're not doing anything. It might feel like you're just... noticing stuff without any clear result.

That's the mapping process.

And if you stick with it—not through discipline, just through curiosity—the Unknown will become Known.

And once it's Known, it becomes Manageable.

And the life you get on the other side of that? At peace instead of at war?

That's the downstream savings.

And it's worth every bit of confusion you pay upfront.

A Final Note on Paying Upfront Costs

I've applied this framework to everything in my life: trading, learning new software, understanding relationships, navigating awareness.

And every single time, the pattern holds:

- Unknown is uncomfortable but temporary
- The upfront cost of mapping is worth it
- Once it's Known, it becomes Manageable
- The downstream savings are exponential

Most people avoid the Unknown phase. They want clear steps. Guaranteed results. No confusion.

But that's not how systems work. That's not how learning works. That's not how awareness works.

You have to be willing to sit in the chaos long enough to map it.

You have to be willing to pay the upfront cost.

And when you do, you don't just learn the thing. You learn *how to learn anything.*

That's the real gift of this framework.

Not just that awareness becomes manageable. But that you develop the capacity to turn any Unknown into Known into Manageable.

And that capacity? That changes everything.

Next: Chapter 4 - "Awareness Begins the Moment You Notice That Something Is Being Noticed"

Watch for more Prairie Light Press Reader Previews

www.ingramcontent.com/pod-product-compliance
Lightning Source LLC
LaVergne TN
LVHW090354240926
848801LV00002B/33
9798996980246